POEMS OF NORTH DEVON

The Coast and Inland · Travelling
Buildings · Seasons and Weather

David Bowen

Decorations by
Anthony Gayton

dbp

DAVID BOWEN PUBLICATIONS
LITTLE TORRINGTON

ISBN 0 9511596 0 7

Printed and bound in Great Britain by A. Wheaton & Co. Ltd in association with CP Graphics of Exeter on behalf of the publishers

DAVID BOWEN PUBLICATIONS

2 Style Park Cottage, Gribble,
Little Torrington, TORRINGTON, Devon
EX38 8PX

CONTENTS

POEMS ABOUT BUILDINGS . . .

POEMS ON THE SEASONS, WEATHER, AND GENERAL EFFECTS . . .

OTHER POEMS . . .

Song of the Western Shore

Rush to the western shore with ceaseless yearning your command,
Desert the clamminess of life with little call for you;
Turn to the unkempt wind and rain, percussive hail, and sleet,
Accept the challenge now, bid welcome to the gale, and grow!

★★★★★

In memory of Kim, my silver poodle, 1972-1986, loving and knowing friend, without whose constant enthusiasm for walks in every part of North Devon that we visited together I would have missed out on so many of my observations

Portrait of Kim by Debbie Peach

NORTH DEVON · Her Coast

The North Devon coast,
A last stopping place,
A wall of grey rock,
White towers on plinths,
Security and
Adventure playground.

Be well assured,
Just half a gale
Lifts a schooner
Clean cut through the
Choppy ocean
And terrifies
The foolhardy.

Onshore breezes
Disturb the coves,
Tussle the grass,
Carry the swell
To the cliff edge,
Undermining
The path above.

The sea falls back,
Foaming, white-flecked,
Teasing the seal,
Laughs at itself
In its music,
Assists the gull
And the wader.

Below the surface
There is no ruler
Unless he is man:–
Man who was greedy,
Not caring to think.
Now he must conserve.

. . . Rurally

A green immersion,
A plan and a dream;
Bold or less rigid
Hills that constantly
Emerge from shadow
And then claim it back.

Pastured cattle,
Exploding crops,
Way-out village
And small hamlet;
Traditional
And persistent
Ways of living.

All features are
Cunningly merged,
Benefitting
Skylark and rook,
Buzzard, and man,
Too, escaping
From boundaries.

Work never stops,
Nor do rivers,
Often looping
To gorge their plains;
Wild flowers are
Resilient
And keen to please.

Hills, if utilised
For towns or main roads,
Cause difficulty,
But upon their crests
Parish churches point
The way we should take.

. . . and in Perpetuity

Ocean and tide and sand
Flow with delight
To satisfy our needs which daily change;
Then, as new year stakes out
Its measured span,
Deep from the faith of well remembered dreams
That ocean flows again.

Rivers and banks and trees
Grow sombre as
Congealing skies regain their mastery;
Ordaining us to wait
With patient hope
Till frothing April quits her wintry games,
And rivers shine again.

Through villages and vales
Find gentle joy,
Replenish failing limbs with sleep and rest;
Then leave, to dwell anew,
Through quiet thought,
In unrecorded hours of time now passed
Those moments linger yet.

Meadows, hedges and hills
Are bright'ning now,
Compelling, irresistibly, our gaze,
Our need for living air,
Desire for song
Of riches given in a theme of praise,
In notes that shall endure.

POEMS OF
THE NORTH DEVON COAST ...
stepping, first, from Cornwall

Budehaven . . .

Rocky shores give way to civilised oasis
Where well placed turfy downs offer the same tempting
Invitation to all but the latest risers
To be about and stimulate the appetite :–

Breakfast cereal will be followed by kipper
Or poached haddock at the two- or three-star hotel
Or the other choice of seaside establishment
Where some windows may face the ocean, or golf course :–

Were it not for the bathers needing the protection
Of twin warning flags and oblongs in black or red
One might think that Bude was pure health resort, harbour
And sleepy canal with no other role to play :–

On the contrary, her seas are as famous and
Furious and Cornish as her many shipwrecks!

. . . Duckpool

Bleak to the end of time, a cove subjugated by its own granite walls, enfeebling
The visitor, too greatly overawed by the massive strength of this place
For his comments to do other than fall as laughable echoes against the darkest
Imaginable rock, its lower face stained indelibly by the sea.
When I came to look at it I had to turn my head. I felt so insignificant.

The Welcombe Valley . . . and the night tide

Lately escaped, but now firmly established,
A fugitive Yorkshire dale that has found rest
On the coast of North Devon, at Welcombe Mouth,
Yet still expending energy, glowering
At those who will never understand, merely
Wishing for convenient relaxation.
Here is hardened peace, roughset vale, transporting
The eye inland from the sea, deviously
Twisting and climbing to where the starkest rock,
Parent to the trembling, oxygenated
Stream, certificates its coolness, purity :—
But now, by the broadening bank, one ponders
On the final award, this green accolade,
Tribute to its passing moments, as the sea
Claims it for its own like goods in a market.

Welcombe's approach, clinging to the valley floor,
Takes the path between wholly dominating
Mountains – shadows falling harshest and barest
When, at last, the coast is reached, influencing
Weather, colour and light, and even one's mood,
Especially near close of day, with the east
Totally mauve, legacy of mid-winter :—
Here the ascending tracks glow in rocky white,
A hillside maze, zig-zag, mask of forked lightning,
Were these not the work of struggling human feet.
Remaining warmth is now well concentrated
On the opposite head, strangely energised
Into brilliant terraces ; and shining,
Too, white rose-cups, fiery campion, gorse and
Foxglove stand sentinel to the copper sea.

★★★

Vanishing, spider-like, into sea-coal,
Welcombe – now invisible ! I, the guest,
Can feel her Mouth reverberating shock :—
No pencil of horizon, flush of sky,
But animated inky tide to test
The new supremacy of cutting rock.

By Hartland Quay

Spirited, bird-wild coast of iron-grey:–
Projecting heads of dead sea monsters,
Reefs of dragons' teeth, salivad
By encircling glassy waves;
Cavorted rock angled
Crazily – once laid
Horizontal,
Now pointing
To the
Sky!

Armies regularly assault the shore:–
Breaker on breaker, reminding one
Of ocean's total mastery
As it undermines the cliffs
Through years unrecorded
Of energy spent.
Results are all
Around – weird!
Come and
Look!

The coast is battered, loose boulders being
Destined to act as giant cannon:–
But in May tall cliffs are lit by
Sponges of sea-pink, soil-less
At the edge – fat, though, and
Shaking all the while
Above the hiss
Of the tide
On calm
Days!

And what of the Elizabethan pier?
It stood proudly for three centuries,
Exporting local malt and corn
And bringing in needed goods:–
Then came the long stiff climb
For all the horses.
The harbour, pierced
By a storm,
Was just
Left!

That steep pyramid, St. Catherine's Tor,
Was once crowned by an ancient chapel,
Yet another miracle, to
Serve as both church and lookout,
Guarding sailors at work
When weather allowed:–
All worshipped God.
Watching eyes
Could pray,
Too!

Nature flows dual-handed in these parts:–
Above Speke's Mill's chasmic waterfall
The turf is trim and garden-like;
And, higher, primitive rock
Has bluebells on its back –
Yet on stormy days
Stoke's church tower
Must observe
Writhing
Hell!

Gallantry Bower

The eye will not part with it:–
The vision is of a wave
Dissected on the angle
Of its tilt, solidified
Into what, in our own terms,
Is permanence – barbaric
In isolation.

It has faced gales since before
We were born and those that we
Do not complain to have missed:–
If it is a warrior
Discounting every bruise,
This is entirely because
They were not noticed.

Its observed ostentation
Is not to be criticised:–
Different to that within
Human beings who, condemned
To show faults weaker than those
Geological, would be
Wise to stay modest.

Here the exhibitionist
Displays totally static
Performance. Theatrical
Behaviour will reveal
Itself, but partly within
The concentration of the
Spellbound admirer –

Who must also pay tribute
To the sea's willing part
In the drama, cajoling
Or winsome, always trying
To capture the entire scene
From the imperturbable
Dominance above.

Clovelly Donkey

Jack has long been exploited by humanity –
Unworthy, one might think, of such an epithet;
Though it was not considered in this light at all,
Merely that a donkey's duty was to work hard,
Haul, and repeatedly carry punishing loads;
And all because he had proved himself capable
And willing, which reminds one about the horse, too!

Just here his life, though above ground, was arduous:–
For him there was no romance going 'up-a-long',
While 'down-a-long' was tough on his calves, like our own;
Yet he, the traditional Clovelly donkey
Was required to burden himself with all the fish
That was caught and to lift it without complaining
Or stopping up the side of a cobbled mountain!

Quite apart from that, he was the general stooge
For delivering the coal or whatever else
Was demanded, which was plenty, whereas, today,
Volunteers with ropes and boards do the 'donkey-work',
While Mister Clever-Clogs, ears constantly tuning
To the conversations of those all around him,
Does what is fitting to preserve his honoured role,

Which may include the sacrifice of leisure time;
For adults cannot resist placing their offspring
Astride his back, and that of his companion, too,
And then it is one long round of photography
While small seats are supported, and commands shouted:–
Human beings, even in our enlightened age
Of donkey welfare, do take some understanding!

Buckish Mills

Your dreams are longer than ours
And scornful of present times
If life, though less gruelling,
Brings you no satisfaction.

Your followers do not doubt
You will insist on clinging
Tenaciously to the signs
Of all your past workaday
Achievements:– your tall lime kiln,
A veritable castle
That dwarfs its near companion,
Needing horses and carts for
Dragging the heavy white stone
All the way to its summit
In readiness for tipping
And firing; then more horses
Taking the fertiliser,
Remedy for acid soil,
To be spread generously
On local or distant farms.

And if tides could be tricky
For the skiffs sailing from Wales
Laden with cargoes of lime,
What about your own great skill:–

Your beachmen launching their boats,
Carvel-built, and as tough as
Appledore yards could make them,
Across the grinding boulders?
You can be proud of your sons
Fishing for herring all night,
And others working without
Any shelter from the storm,
As waves regained their control
Over your only defence;
But then a low breakwater
Was better than none at all:–
'Improvise, improvise,' that
Constant cry of your sailors,
Always served them in good stead
And kept them resilient.

Do you regret time's passing,
The end of your usefulness,
Or even feel deserted
Without justification?

Your successors now talk of
'Bucks' Mills, instead of the old –
May I call it truculent? –
Handle of 'Buckish', shortened,
Inevitably, to 'Bucksh',
Which gave to the mapmakers,
But, at first, the Post Office,
The excuse to prune it down
Still further, perhaps not then
As comment on your lifestyle,
But yet barely concealing
A hint of degradation,
That later change of status
As fewer men put to sea:–
When gentler times brought decay
The camaradie took flight.

And while you attempt to fit
Into the world of today
And feel the lighter footsteps
From visiting inspections
With calm equanimity,
Yet there are some even now
Who wonder if this new role
Of yours is appropriate?

Old Times at Appledore

Talk about boats – they grow here like flowers!
Those that are not floating are in courtyards
And gardens of this virtual island,
While, in tidal creeks, vessels are being
Built with beautiful elliptical sterns
By Mr. Westacott. Few are tougher
Than his, you can be sure!

It will take time to see all that is made
Here in Appledore. Nothing is standard:–
There are salmon boats, lifeboats, polaccas
To save tacking when sailing into this
Port across the treacherous sandy banks
Of the Bar, as well as ferries, smacks, brigs,
Trawling skiffs and schooners!

But if your square rigger has faced high seas
In far oceans, its copper sheathing worn
And ragged, seams open and leaky, then
Mr. Yeo of Richmond Dock is your man.
Think of it:– his father went penniless
To Prince Edward Isle, working in timber,
And returned a magnate!

Where we stand now a vessel's sharp boom is
Nearly touching the houses opposite,
And along the New Quay sail-lofts, workshops,
Anchors, shackles, dead-eyes, blocks and wire strops
Lead one to cottages and fishing gear
In wide variety, and where ladders
Are all spiked to the cliff!

Little wonder the Excise finds its life
Is unrewarding:– contraband landed
Quickly in a hundred places at once
Goes straight into villagers' back gardens;
But in any case the Appledore men
Are no easy catch – they are hard bitten
And eat missionaries!

The boys, too, are living legends, bootless,
In blue jerseys and calling men 'Cap'n'
While still learning to be pirates:– they use
Nets, climb aboard ships, leap on to rocks,
And launch boats, which is what one would expect
In a place that is maritime down to
Its very last shaving!

Isle of Brigands

Eerily shaped, a distant grey Atlantic squall
Against the horizon, or linked with banks of fog,
Innocently soft, pliable, as they extend
Oceanwards, creators of insecurity:—

Isolated, even when the sky shows promise,
By majestic seas, pale waves turning to sapphire,
Detonating as they flail the granite defence
That will not shift, splitting acrimoniously:—

Cast for adventure by strategic position,
Insistent on following its own destiny
While offering continued opportunity
For harassment along a main shipping channel:—

Fortified by a succession of overlords
Endowed by medieval kings who, in return,
Were ignored, or disowned, or became the victims
Of treasonable plots, with their own land plundered:—

Accommodating as a base for privateers
When government gladly connived in extensive
Operations by those who were willing to fight
The Queen's enemies, keeping the spoils for themselves:—

Then acquiring an outrageous reputation
As buccaneering beacon, with glittering gains
To be realised by attacks on lawful ships,
But, in turn, raided by foraging French and Turks:—

Attracting entrepreneurs for the purpose of
Indulging in highly profitable ventures,
Such as illegal dealing in prisoners by
Thomas Benson, M.P., who then built walls for him:–

Visited repeatedly for buried treasure,
Which does not appear to exist in spite of the
Armada galleon and other laden ships
That met their doom – and for its wonderful bird life:–

Puffins, that gave their name to this isle, shearwaters,
Guillemots, kittiwakes, razorbills and falcons,
But the geese have disappeared, and the goats hide well,
Still remembering the days when they were hunted

Among the rocks – where the impact is unfailing,
Always grotesque, revealing nature as sculptress
In a land whose earliest times are still clouded
By doubts and conjectures, so fitting for Lundy!

Mosaic for Sale

The product before me is an unusual one,
A prototype flooring in rich mosaic pattern
With immediate commercial potential that is
So great a variation on all other known blocks
Or tiles as to arouse immediate attention:–
In beauty and richness it surpasses all of these.

Initially it could serve as a one-off venture
For a wealthy businessman in his own mansion
Willing to undertake all the development costs
In this respect, so that, possessing the floor himself,
He would derive not just personal satisfaction
From its use but also the approbation of friends.

Later, as soon as he was willing to relinquish
Exclusivity, he could regain the net outlay,
As well as benefit from the accruing profits,
With the fabrication made widely available
And of course rationalised:– the present somewhat too
Gravelly texture could be improved by machining.

The original, that to date has only appeared
As a fortuitous occurrence – suggesting that
It would be unwise to delay sponsorship too long
In view of competition that is bound to arise –
Can be viewed to advantage any day at Braunton
Without prior appointment: the inventor is dead:–

Or rather, I should explain that many were involved
In the making of this unusual strip design,
And that was some eight hundred years ago, I believe;
But very long after the Taw and Torridge rivers
Had laid the groundwork, possibly a hundred thousand
Years before the Braunton Great Field began to take shape.

Bondi Blue, if you please

Woolacombe, open jaw to Atlantic!
Boxing gloves, your waves, real knock-you-overs! –
I just wonder, are you ever static?
I'm not suggesting you copy Dover's
Oily calm at the mere drop of a hat,
And pure chalk cliff so tall:– yet unlike *your*
Weathered old head and all your sand – so that,
Well, you've plenty over to share with poor
Neighbour Putsb'rough, which takes it, anyway!

So, if you're still feeling obliging, then,
There's a task for you that I'm bound to set:–
Now your clifftops are green as a garden,
And your sea inspiring, wide awake, yet
I have to say you're not wholly Bondi.
In Devon, yes, you're a legend, I know –
Long rolling breakers and wild winds that try
To keep them coming; but why don't you blow
Off some of your cloud? We can't surf in that!

Combe Martin-Style

Demolish all false notions:– what, agreed, is north
On the map is distinctly southern in flavour,
Gripping its coves and 'arbor – one apostrophe! –
And just a touch of Mediterranean, and

A lesson in geography on arriving,
Therefore. But remember (if you can) centuries
Of mining as well:– silver, lead, iron ore, and,
Would you believe, manganese? – but principally

The first of these:– could London's lord mayor, with name
Of Martyn, no less, be denied his silver bowl?
But this was unsolicited and long before
The armies of Plantagenet kings dug deeply.

Then should you determine, lest you acquire a stoop,
To take a mental break from what is buried, look
To those riches well above ground, to rarity,
In fact:– intricate, delicate and masterful

St. Peter ad Vincula, built not in local
Material quarried throughout North Devon and
Never to be thought of as lowly or dowdy:–
Just that this great church in rose-coloured sandstone can

Be compared with mightier structures, even if
She does not dominate, leading one to suppose
That Combe Martin began her life differently
From other places and, quite unlike Ilfracombe,

She evolved from way up in her valley:– the sea
Came later – for Victorians. Pause, if you will,
To imagine the make-haste among those inclined
To improve their view, or their health, or both at once,

And see this ocean annexation as it was,
Not as a blueprint centuries late. Obvious
Courses are always those viewed retrospectively! –
But ask why no real 'town' exists to match the hall,

And then accept what is perfectly straightforward:–
Grand merger of the principal streets into one;
Or did they first do battle? But perhaps it is
Merely a quibble to become preoccupied

With such matters of status:– one should be thankful
Indeed that this is no circulatory route
Of an adapted medieval town centre
Where traffic imitates the art of a fairground! –

A situation that most drivers would avoid
If they could, and in this context, as in others,
Long live Combe Martin's idiosyncratic traits
That laughingly amend established precedents:–

Not least that her well-favoured tropicality
Has been harnessed without a single discordant
Edifice to confront the sky with its concrete,
Thanks to modest intimidation by 'hangmen'

Whose function is now merely supervisory! –
Then there are legends of strawberried mountainsides,
And hemp, and strange wild pear, within this quiet place
Whose toes may touch the sea while head accepts the cloud.

Ilfracombe New Town

Life will continue when the chapel goes from Lantern Hill:–
Pretty, it is; but not in keeping with an ocean port
In modern Ilfracombe. First, the marina, cargo next
Within the enclave for extended docks; containers, too,
Ensure, together, work for local men with expertise
And scope for builders, civil engineers and allied trades.

At anchorage, with navigation well below ideal,
Water sufficient for a sheltered deep sea terminal
Necessitates explosive charges under surplus rock.
The Navy's role is understood:– thus, by the outer base,
Equipment there will monitor approaching submarines
And aid the civil power to check irregularities.

Depressing news! Latest marine inspection shows a snag:–
To dynamite at Hillsborough may cavernise the Head –
Or so consultants say – it could collapse quite suddenly!
But extra houses may be built with funds available;
The government agrees it will expand the town to meet
An estimated future need, including industry.

They've allocated Chambercombe for manufacturing:–
Heavier goods will then have access all way by main road,
With shift of population up to new-design estates
And qualified domestic use of Killacleave, or near –
The manor house is safe:– facilities will be installed
With eye to education, art, and public gallery.

A park-like ambience will be preserved at Bicclescombe:–
Certainly, that is, within the total infrastructure
Plan for town hall complex, offices and retail outlets,
All glass-domed, with restaurants, to be incorporated
In latest architectural image! Be sure of this,
The project must be forward-looking, and significant.

In time the present High Street shops will be superfluous:–
Walking a mile with heavy goods, and hat, and overcoat
Is retrograde. Instead, envisage fine raised promenade
Replacing lower wandering from Wilder Road, which then
Accepts its designated role of portway if, in time,
The dockland scheme is passed. Together, all adjust, or fail!

When the widened Highfield traffic flow has been approved, new
Seafront priorities, and compensation for hotels –
Should income levels fall appreciably below the norm –
Can be discussed, Pavilion resited, beach made safe:–
Tall concrete artefacts from former High Street down to shore
Will prove a major tourist asset and attract a grant.

Programmes tailored to stimulate the town's employment can
Each, individually, enhance the basic concept;
But, first, the overflow from other towns must be assessed –
While need to satisfy the old-time conservationist
Can take the form, museum-based, and working-model-wise,
Of Ilfracombe in former times when people were deprived.

Woody Bay

Technically escarpment, a mountain elbow rising from the sea,
But surely a stone crocodile with his solid neck in the water,
And, while motionless, forming the western promontory of this bay;
Then, next to it, his resting claw, a stubby foothill that takes the eye
Ever aloft to survey the whole thousand feet of tangled green oak –
Stunted trees that cling to the rock as ivy does to a garden wall:–
Ancient roots are wonderful adhesives against anything the wind
Contrives; and with the waxiness of each maturing summer mantle
When one would expect such a towering northern barrier to light
To produce the darkest shade imaginable – though not unwelcome
If the shore is to cherish its own secretive and magical role –
The shadows are seldom black but, rather, mellowed as if by a veil,
A vapour created by the ceaseless fretting of obstructed waves
Or derived from the colours of the bay, seldom completely pallid,
Remitting its blue or green, or a penetrating misty silver.

Yet that stone crocodile, inert and toothless though he may be, still hints
Of danger, of the upturned hospitality of sculptured fjords,
Or their near equivalent, open to the Atlantic and her whims –
Storms that count as little more than myth in the lassitude of summer
But are always booked to return:– witness the remnants of the pier
Designed by optimistic Victorians who had earmarked this bay
For the development of a high-class resort. It soon protested,
Being far too rumbustuous to become lackey to ambition.

Spirit of Lynmouth

I ask you for the thoughts of folk who come
To Lynmouth sea, relax, breathe deep her air,
Caress with hand her parapets of stone,
Approve her festive mode – exploring, peer
Through Rhennish eye, take colour photographs,
Discuss the recommended stops, vote for
An hour or two at Watersmeet, or else
The valley of the Doones, or Badgworthy . . .

Please tell me what is in the hearts of those
Who, coastal-bound, pause by the river bridge,
The pebbly West Lyn course in scented flow –
Delight of car-borne confraternity! –
To find they are reluctant to move on;
But others, reaching Lynton far above,
Hydraulically, by atmospheric rail,
May climb with goat for rocky vantage point . . .

I'm curious, they've come to 'ground' again
To sit, sedately, by a well known inn,
Consume proprietory brands of fizz,
Yet not ignore the local dairy ice,
Try on a feast of brightly coloured hats
That shield the eyes, but catch the roving ones,
Then think of souvenirs for mantel-shelf
Or, child-directed, bouncing balls or kites . . .

So, now, what is it motivates all those who share
    the Lynmouth heritage –
The coast, with Switzerland and Lorna Doone,
    kaleidoscope of sea and moor?
Would any port suffice? In that event there was
    no flood, no night of fear.

Perspective from Countisbury

Think, first, of a china cup and absorb the glaze
With your eyes, feel this flatness
Or the wide-angled crinkly silver or pewter
Taken from a roll of foil;
With cloud absent sun and haze determine colour,
These two are good companions,
Or if the wind blows intemperately up here,
It does occasionally,
From this well-charted height of Countisbury Hill,
Which once you vowed you would walk,
No real waves are seen, only small white polka dots
As on a heatproof worktop.

That eye of yours insists on finding an object,
Which is why you came up here,
Although when it does it creates controversy,
Even with binoculars;
For the mast that has just yielded to your focus,
And led to consultation,
Is causing you a degree of uncertainty,
Though, soon, all may be revealed:–
It could belong to a dinghy – or a frigate.
Take your choice and announce it,
Unless you are looking at a supertanker,
Making allowance for smoke.

Perhaps you should conveniently lose the ship
Passing the binoculars?

POEMS OF WOOD,
VALLEY, AND MOOR....
with a finale from Somerset

Hooper's Water
(Yeo Vale, near Bideford)

I would spend a minute by the small stone bridge,
No longer than that, for I had already
Extended my walk, admired a massive oak,
And stood to inspect a creditable feat
By house martins manoevering on wires,
'Dressing by numbers' and taking a roll call,
But of all natural elements water
Is surely the most compelling, insisting
On sharing its timeless affairs; and bridges,
When not archways of effervescence, treacly
Or sparkling, will gladly oversee the pure
Crochetwork of currents on debris and stone.

A minute, did I say? What would I have done
With any saved by turning my head and steps
When my allotted and quite inadequate
Span of time had elapsed? The point, however,
Is academic. On hearing the gushing
Of water as though from the sluice of a lock,
I felt the pull of that bridge, mind upon mind,
Revealing, as it happened, not the main flow
Of clear river spreading from beneath my feet,
But a small culvert emptying its contents
To one side, the brash genie of the valley
Who had it in mind to usurp the whole stream.

I say again, time's web makes fools of us all
Until we no longer seek to oppose it,
And those who passed by me wishing the road straight
Have much to learn, as did my bustling culvert
Intent on a sprint that he could not maintain
For long and the river would not countenance:–
Sideflows in today's softly gouging water
Were not to be allowed to dictate its pace,
Which was lackadaisical, Septemberish,
Easily measured by the transport of leaves
That appeared from under the bridge, one by one,
Fearless entrepreneurs in their new harsh world.

How would these voyagers fare in the rapids
That lay ahead:— rapids, that is, to such frail
Initiates torn by the recent high winds
Before their time, and still in brilliant green?
Possibly they had already encountered
Turbulent currents before reaching the bridge,
And now they approached liquid Gothic vaulting,
The pattern created by a large stone block
On the bed, acting as a conning tower:—
I expected them to be harshly treated
For being so daring, but I was proved wrong.
They dissected the waves, bobbing as they went.

I noticed that the leaves remained in mid-stream,
Avoiding the shallow pool at whose clear base
Well embedded stones were as neat and varied
As the wall of a flinted country cottage,
While representing a mirrorback. Above,
The surface of glass was engraved by a wind
Unfelt on the face, as light as a whisper,
Yet creating delicate ripples, eyebrows
Lying on the water, the perfect foil
To the stream's mysterious distant plunging
Where, anticipating the bend, it is forced
Into crests, bright galloping steeds on the rocks!

Castle Hill, Torrington

Perched, jauntily, on a precipice
Where Devon copies Jerusalem
Or the greener heights of Tuscany
But with an added luxuriance,
Recreating or else reminding
One of locations hot to the touch:–
Torrington, Italianate, then,
With her panoramic open side,
Is an approved pedestrianised
Viewpoint for all who care to wander
The convenient yards from the shops
To take an imaginary leap
Or, more likely, sit, or walk one's dog.

Here the river curves as an archway
Between two competing focal points:–
From the sleeping stones of Town Mills bridge
To where the other, at Taddiport,
A nucleus of activity
Under hovering banners of smoke
And an aroma of milk powder,
Finds relief in new suburbia –
For surrounding bungalows enchant,
When mingled with field, river and bridge,
To be perceived as reflecting stones
Coalescing with nature herself
In the clear sharp light of afternoon.

Replacing the shaded higher course
Of the Torridge's wooded canyon,
Detect, at Castle Hill, an ordered
Watery floor, compliant grandeur,
Even elegance, especially

The parquetry of hedge and meadow
That rises from the opposite bank,
Encouraging the mind to ponder
On how it was formed. Was symmetry
A deliberate aim? Maybe not,
But all the conveniently spaced
Municipal seats in tasteful green
Show the hand of the imitator.

I am not certain, I just enquire,
Am I ungrateful if I complain
When ruggedness is tamed to provide
A public garden amenity,
A place to skim the daily tabloids
And then, as an antidote, saunter
As far as the monument and back
Before returning to town or car –
Except that what comes too easily,
That can be consumed like potato
Or petrol, without much prior thought
Or the aching that patience creates,
May be valued less on that account?

And does that magnificent outlook
With its ordered life and patronage
Lose quality by being constrained
For us, like a tiger in a zoo?

Shadow on Hatherleigh

Stop, stranger, you are at Hatherleigh,
Under the grey mountain shadow:–
I, though, am that stranger, as I tap
My pocket, feel for documents
Before remembering that I dream
Of an age that has not occurred;
For there is no one at this border
Who will question what I carry
Or give instruction for my journey:–
An indelible approval
Is the encouragement that I need
On the clean page of my passport!

I have no wish to desert this town
Without the adequate backing
To link me firmly with the green land
And my right to return to it:—
I am heading for uncharted bog
And unlimited piles of stone
In a wilderness where few have gone
Singly, hoping that religion
Has found its way into these mountains:—
Or do groups of vengeful natives
Wait within their smoke-filled dripping caves
To pursue their plunder — and me?

In the present volatile winter
I immediately discount
Yes Tor's beguiling signals, her face
Now fresh, though clothed in irony,
Is contracted to the gods of storm
And the spirits that they release:—
Isolating blizzard, avalanche,
Torrents to roar through the gorges,
An incandescent, violet sky,
Or creeping coils of mist joining
Together in a conspiracy
To obliterate all mankind!

Track above Beaford Bridge

Above the bridge a wooded brilliance, a rising stage
Shamed until now by daylight gloom, but emerald today:–
That familiar prodding blade of ocean's nor'west breeze,
Chilling invader, canceller of spring, is in retreat,
And where we mount the track, then pause for breath, enchantment reigns –
Call to the gods for company! No, wait, we merely know
The sheltered yards, the bracken path; our paradise revealed
Through sundrawn bluebell scent may yet be limited in time.

Does fear of loss within mere thought create unwanted change? –
Cloud clinging to the unseen hilltop is some mind indeed
To activate the creeping hands of grey on lower banks,
While we ourselves deplore the shadow, warning cries from birds,
And gusts through tops of swaying fir descending to our heads.

These forces are in turn reversible, they will not stay:–
Look to the curling river's farthest edge where brighter scales
Begin anew and, moving fast, rekindle light and form
In upward spiral, welcoming and honouring our faith.

Country Retirement near Winkleigh

Ne'er a thousand but two hundred years ago
Savages roamed in these parts;
And there's wildness still among homes built of clay,
And town is a long way off.

Find your cottage but be resolutely sure
Contentment comes from your peace;
For there's no rush now in the long days of thought
And sight of everything green.

Make close friendships, not commanding your estate,
Tolerant be where you can;
That old urban dash when you're breathless and late
Is game for a race apart.

Keep your transport, but a saunter here and there,
Footing it hard all the way,
Is the cure-all now for the things that are past
And zest for the life today.

A group of people well-known for their wild behaviour
were once known in central North Devon,
hence the reference to the North Devon savages,
in the first stanza.

Bursdon Moor

Hauntingly bleak, no more barren
Elevation, surely, than this,
Except for the highest moorland –
A hurricane for all on top
In stage-coach days. Hear the wind moan!

From Clovelly one way, or back
From Welcombe Cross, more modern man
Still feels the gusts as he urges
Himself to clear the serpent road
Where gradients are misleading!

Growth is limited to the tough
Genuflecting grasses – heather,
Gorse and, lately, conifer, but
Imagine one's garden here and
Writing to experts for advice!

The open moor, rusty or grey,
Marks the mind, may even depress
Or suggest evil spirits, when
Substitution of shelter for
Bog is a recipe for fear!

But wait for a change in weather
When raw desolation gives way
To the safety of bright sunshine
Under the clearest horizon –
And then be glad you have escaped!

Across the Torridge Vale to Exmoor

Far-flung Exmoor, reposing whaleback, but untiring,
Holds me motionless, and calls. Her semi-camouflage
Of purple haze enhances, does not mar, her outline
Or her influence. Long has she counted Torridge Vale,
Over which I stand behind a wooden five-barred gate,
Among her charges – a rolling countryside that now
Presents only velvet worthy of a Devon spring
And attracts the eye with measured calm complexity.

Below the moorland edge the skyline, cut like metal,
Grows individual, or groups of, thin stalky trees,
Each with miniature umbrella held to cloudbase,
While a lower ridge on static ocean wave in green
Takes highway, cotton-frail, from Torrington to Beaford,
So obligingly revealed by tops of vehicles:–
These crawling beads are, like the numerous dotted sheep,
Contributing allegiance to the chequered scene, with
Mirrored farm-roofs, trees, hedges and copses all flexing
The eye this way and that. A further point of focus
Is the lower curve of rounded hills like half-clenched fists
And from whose centre a forest of intriguing shape,
Deciduous Brazil nut, tips to where the unseen
River, carving out a horse-shoe, adds to mystery;
Then, from the wood, the broadening pasture, firm and steep,
Rises to my gate, the chaffinches, and almond air.

More than that, the whole land is honeycombed by sunlight
Mixed with shadow, an ever-creeping mould, puppet-stringed
By gathering cloudy fragments in the clean-washed sky,
So that each ray, valley-bound, is received by meadow
Like a lantern, yet as though the light came from within.
North Devon, forgetful of past scars, is in her prime.

Ghost on the A377

Some might declare that a place
Notable mainly
For its level crossing,
Gloomy state forest
And little else
Lacks interest.

That is a material
Judgment, however,
With no application
For the keen student
Of history
Or atmosphere

Such as myself, a passing
Observer only,
But one who will not be
Likely to forget
An incident
In this forest

When calmly driving through the
Dusk of Eggesford,
Knowing nothing about
Its former mansion
With its secrets
And memories

And just anxious to get home:–
The year in question
Was 1959,
When, for the first time
I can recall,
I met a ghost.

I admit to feeling scared
Despite, one would think,
Being completely safe
Inside a motor
Vehicle that
Was still moving

Because it was at that point
A fateful breakdown
Seemed inevitable,
Leaving me alone
Without any
Proper defence.

The spectre was tall and gaunt,
Bald, wearing a mask
Concealing half his face,
Moving with a slow
Sinister gait –
Ignoring me!

Had I been within the wood
When thus confronted
By the apparition,
With no visible
Means of escape,
I would have died!

But it was still unnerving
Keeping to my course
Along the forest road,
With night approaching
Fast – somewhat too
Fast for comfort!

Call this a paranormal
Experience if
You will, though I confess
That some weeks later
Exactly the
Same thing happened.

I will accept your verdict,
But I now suspect
Anyone passing by
In the grey twilight
As though I, not
He, were the ghost!

Challacombe

Conceived in cloud,
Highborn and free;
Your father, moor,
Your mother, hill
To deep valley
Floor.

No sweeter place
In Devon birth;
So placid, mild,
Inviting sleep
Under contours
Wild.

I miss your peace –
May I return?
Will there be room?
Floating, my *thoughts*
Will find Challa-
Combe.

Evening on Dunkery

Warmth ebbing in the mist, houses snuffed out
Like candles, the sun-shafted trunks within
Horner's green bowl accepting solid black,
And stone beacon-hearths, my near companions,
Becoming airborne through isolation:–
Thus does the day's unchallenged overlord
Renounce his vast territorial claims –
Wales has long vanished, and all sheep wishing
To stay white must remain in the foreground;
They are no longer promoted by blue
From the Channel's flattened ultramarine
Or the higher one crammed with soft fleeces
That imitate pastoral contentment.

At my feet the heather in pillowed clumps,
Still fragrant in the night's humidity,
Blooms to perfection in the slanting sun:–
The radiance is deepset, passionate,
A floodlit cathedral-like purple veil
Created by vegetable matter
Into the most vivid Gothic grandeur;
And while the day's clarity still remains
Sharp in the eye, and will not be severed,
Neither will oncoming night blot it out,
But, rather, allow time to recreate
Its past exuberance to meet the morn,
An eternal arch that cannot be lost.

Then, with the sudden drop of the day's breeze
And all shadows lovingly communing,
The pattern of observation changes:–
Sounds become more meaningful, personal,
And will be stored in the mind for ever,
Like the single fading cry of a bird
Followed by an overwhelming stillness

That, in a strange way, resembles the wind
Of the afternoon: yet nothing moves now
Apart from pure thought, so gently whispered
But carried effortlessly over hills,
Uniting the meditation of years
To this one moment here at Dunkery.

POEMS OF TRAVELLING....

Begins North Devon Where?

Begins North Devon where?
After Wiveliscombe,
I always used to think;
At the precise moment
That the unstowed luggage
Left the rear of the car
To assault the front seat
Passenger on the head:–
My driver at the time
Had not been accustomed
To the hairpin corners
That were symptomatic,
Or so we all agreed,
Of ancient North Devon,
Not gentle Somerset.

At St David's station
Passengers are summoned
Into physical jerks
By crackling loudspeakers:–
They obey the command
To hurry to the train
Then destined for the North,
Which means ascending steps
To a metal gantry,
And then rushing, tripping
To the impatient coach
That, as soon as it starts,
Is more for sauntering
Than attempting to break
Its previous record.

Where you feel North Devon
There, surely, it must be,
And particularly
Along that high outback
Above Wiveliscombe;
Or even taking tea
With scones and clotted cream,
Served with great dignity
And accompanying
British Rail silverware
At Waterloo Station
After boarding the train.
Of course, it's different
Nowadays, and they make
One change at Exeter.

Very soon North Devon
Will suddenly appear,
Holidaymakers think;
So where is Crediton,
They demand, and is there
A town and cathedral
Around Morchard Bishop?
Forests and prairie-like
Farms all rush past the eye,
And with the river scenes
Of the languid Taw Vale
Central Devon is left
Behind:– yet none of this
Resembles the brochure.
Umberleigh is not coast!

The explanation is
That seasonal commerce
Ignores the inner face
Of North Devon:– Beaches
Are where profits are made!
Yet they remain concealed,
Even at Barnstaple,
For railway travellers
Nowadays, which all helps
To whet the appetite:–
When new and faster roads
Bring the shoreline closer,
One will arrive quickly
At Hartland or Lynton,
But miss some of the fun.

For myself I confess
That I like North Devon
To begin remotely;
Then it becomes sharper
As the route continues.
Once I even tracked it
At Slough, which is well known
For having its own scent:–
A new notice appeared,
Unfamiliar as
It was welcome to me –
'To the South-West,' it read,
And I took it, feeling
Quite at home, but in fact
I was bound for Portsmouth!

Byroad to Barnstaple

There are roads that get one there quickly,
Or this is at least the intention,
But well in advance of the river
The choice is to leave the car and walk
Or drive at less than the speed of feet:—
Yet, taken in time, there are byroads
That, with the aid of an ordnance map
And a competent navigator,
Can be harnessed and are guaranteed
To provide scenic and other joys.

An alternative to map-reading
Is the often-tried remedy of
Testing the knowledge of an expert
And obliging native who appears
To have no urgent task to perform
Other than to answer one's questions
And who, both by word and sleight of hand,
Will indicate relevant details
In a manner that is befitting
To this county of cider and cream.

The only trouble with directions
Given by such lucid amateurs
Is their very comprehensiveness,
But it is difficult to dampen
The zeal of an accommodating
Bystander once it has been harnessed:—
One must therefore go along with him
And picture the cottage with the wall
Where there is a turning to the right
Yet accept the need to keep straight on

And ignore the signpost at that point,
Although both routes lead to Barnstaple
In the end. Yes, yes, you bet they do,
But what about the urgent shopping?
And how can one possibly express

Gratitude with any conviction
When all that is needed is surely
The briefest of general summaries,
Not to have the mind utterly fogged
Before one has grasped the half of it.

Perhaps one may cherish the hope that
All will be revealed as the journey
Unfolds, that the sage's directions
Will cease to be an obscure jumble
And become meaningful? Certainly
There is little cause to believe that
One's fellow travellers will reveal
A better grasp of what has been said,
For if they did hear anything, *now*
Is the time for them to disagree

Among themselves with voices all raised:–
Pity the poor driver, who pauses
For the very briefest of moments,
Half inclined to ask the patient guide,
Who is aware that all is not well,
To begin at the beginning, but,
This time, without the embellishments,
Just to give the memory a chance.
No, that would be quite unthinkable,
The man has clearly put himself out

To be helpful, and with spirit crushed
The second time round would be hasty,
A product of his embarrassment
And, doubtless, incomprehensible
On that account. There would be no gain
From the exercise. The obvious
Course must be to rely on the map
As a sole aid, and pay no regard
To the cottage on top of the hill
With the stone wall and the yellow gate

Where there is a signpost, or was it
Just a turning? It must be ignored
Because if there is a multitude
Of cottages with stone walls and gates
Of the correct colour the results
Are likely to be quite disastrous,
And the shopping will never get done.
Such clues are not for the likes of us
Who would fall victim to the mishap
Of ending up, lost, in some backyard.

At this point one might need to confess
That the map inside the car glove box
Is not the answer to the problem,
And although signposts proliferate
Few if any of them wish to know
Where Barnstaple is situated,
While to rely on intuition
Increases mileage and demonstrates
How easy it is in North Devon
To discover that the Earth is round!

A devious but rapid route to Mortehoe, or anywhere else! ....

Evocative, aren't they, those little situations that send the mind positively reeling:–
A few notes of music can do it, or a word or two that one picks up as one goes along,
Like when I met my farming neighbour who was arguing with his wife and *would* call her 'Mother'! –
Or was it that gust of wind that blew up the bedroom curtain like a sail and broke the best vase? –
Or that idiotic man blasting his car horn by way of greeting:– I could do without that
Exhibitionist! Well, I suppose it might just have been him, or else the farmer, or that tune –
Schubert, I think it was – lovely man, Schubert, so lyrical – anyway, as I was saying,
Once you get something to remind you of something else, there's no stopping it at all, like a cloud:–
Indeed, I saw a really marvellous cloud the other day, with a vast blue patch, like a lake:–
Still, there we are, it was just that something reminded me of Mortehoe – *beautiful* holiday! –
I suppose it might have been the music, or poor 'Mother', or the vase – but I'll never know now!

. . . and foolproof directions to Little Torrington

You know the way? First, cross the narrow bridge, say
'Torrington, good-bye; next, Little Torrington':–
Type of highway? Not mountainous, but hilly.
Undulating? Hardly:– Yours is upward bound
Countryside, one long ascent too lovely to
Intimidate – yes, that is too strong a word:–
Compensate, that's what the journey does, I think:–
What I mean is, rising, you traverse fine woods.
Progression is by way of sinuous curve:–
'Never-be-good' is what they call one thicket.
Elucidate? No one told me why that is:–
Slippery and steep, maybe; but, no matter,
Remain on course, look straight ahead, and, mind, now,
It's sharp left when you reach the top, or miss it!

POEMS ABOUT BUILDINGS...

Anglican Trinity

A trinity of pinnacled towers,
Exemplary high-rise architecture,
That, grey-coated, tell of distant years,
And in whiteness reflect sun, not temper;
Each in its role of lonely signaller,
Beckoning, with hilltop message, always
'Be of good cheer,' Christ has 'overcome
The world,' but incorporating so well
The tenets of their historical links,
And even the purpose of history,
That while successive years have meaning,
Individual ones may produce fads
And divergences that add up to nought:–
Accumulated faith will not splinter!

From a favoured open point on the downs
Approach these individual towers,
Parkham, Alwington, and Buckland Brewer,
Just one of which is handsomely tapered:–
Witness the achievements of mere stone blocks,
From battlements, belfries and buttresses
To marked string courses, gargoyles for water
That might annoint an unwilling bare head;
And knitting the three lofty parishes
So effectively together that what
Lies between them may well go unnoticed:–
While the towers praise God, the eyes praise *them* –
Such is their power, their abiding charm,
The welcome spirit of the C. of E.

The Chapel makes its Mark

If not small and cottagey, an auditory
Rural shell that is pleasing, though bare, to the eye,
Then rococo Gothic, or porticoed splendour
Of neo-Classical design – shining High Street
Masonry, conspicuous, tall and imposing,
Vying with Anglican grandeur, and embracing
A clock against its tower, emphasising that
The use of time is crucial whilst we are on earth;
Or else the 'wayside pulpit' does so, iron-railed.

For centuries only bare crosses, then churches
Marked the landscape. New innovatory chapels
Were not envisaged; yet nowadays their presence,
Reinforced by singing that filters or echoes
Across the valley seems to belong to the same
Mists of time, except that customs change:– the head bends
Rather than the knees. But if the farmer praying
Next to you sold you a cow low in butterfat,
Then, just as in church, you must stifle wicked thoughts!

Traditional Dwelling

All the cosiness of cob and thatch,
Walls bulging like sacks, but they've been like it for years;
Ancient beams hold them firmly in place:–
They came from an old brigantine, the owner insists;
He even knew who the captain was:–
Anyway, they are used to riding out hurricanes.
Of course, there are a few draughty spots,
And the whole place is crying out for double glazing –
The salesmen will call any minute –
While a surveyor could make a wonderful report
On it, running to fifteen pages:–
But with a reassuring flicker from the stone hearth
Lighting up the copper ornaments
And throwing grotesque shadows on to the farthest wall,
Where the plaster has long been flaking,
Modern standards can be safely postponed – let us say,
For another hundred years or so.

Solitary House of Stone
(North of Winswell, near Torrington)

Either one fails to notice it
Or it is irresistible,
Denying the claims of others
That are too néar to please the eye
Or cringe half-buried in the mist:–
A single gaunt house, now lifeless,
Where the uplands whistle with cold
And winter snows lie undisturbed,
And where, at night, you plod your way
Through the tedious rough climate
Until hospitality looms
Bravely, as a ship beams into
The surrounding void with its lone
Challenging island of pure gold.

My solitary house of stone,
Tall, limey-green, and windowless
In its gabled austerity,
Is causing me to speculate
Wildly. It is rooted in time:–
It has gone its own way, and will
See that its occupants conform
To what has been. It shows contempt
For modern fads by the simple
Expedient of not knowing
Anything about them. It sees
Only the empty road scaling
The hill beyond like a ladder
That has been anchored to the sky.

There must be an inn somewhere

Do not catch cold on a hostile shore,
Too much walking is bad for the feet;
You may await the better weather
Warming up a comfortable seat:—
Why waste your time in prickly heather
When all you need is an open door?

POEMS ON THE SEASONS, WEATHER, AND GENERAL EFFECTS...

Torridgeside in April

Sky in torment, a vast encumbrance,
Descending to the ground, sweeping it,
Swamping it, adding unwelcome lakes
That are constantly whipped into swathes,
Changing the face of the ordnance map,
Making it up to date and truthful;
But it is the fervour of this gale
Which is troublesome, much like the last,
Refusing to stay over Greenland
Where it first sprang into active life –
That, for the record, two days' ago –
Except now it is April, surely
No longer a winter playground, as
The shoreline continues to thunder,
Breakers convert concrete to rubble,
Even copses inland are roaring,
And gulls, willingly or otherwise,
Become the flying ammunition
Of this latest freak from the Arctic.

The Sight of Spring

(Around the Taw and Torridge and their tributary valleys)

Healing gifts in the wake of surprises
When winter is allowed its final say
And is scolded for being untimely:—
But, always, the sun shows full repentance
And is never more inspiring than when
It discovers *this* year's first green mantle
Framed by a consolidating yellow,
The gift of primroses, celandine and
Dandelions posing as marigolds,
Though daffodils seem to bring forth late gales.
More warmth achieves a multi-coloured blur,
But cherry claims its own pink horizon,
While bluebell comes to soften campion,
The spirit of valleys and riverbanks,
And orchids show their purple brilliance
After the snows of blackthorn and hawthorn,
Reinforced by floating cumulus cloud:—
Then blue is in command, with haze and heat
That, in late May, seem out of character;
As, too, does the first major thunderstorm.
Soon the heat may return, but not the spring.

Invocation to Summer
(Between Dartmoor and the North Devon Coast)

Is it June at last? No matter,
The thatch is warm now
At the moor's edge.

I test the burning summer sand
And its antidote
In tepid pools,

Or bathe in the last liquid heat
Of the grey night breeze
Along the lane:–

I will find a patch of firm grass
Fresh as granary
Or cottage wall,

Honeysuckle in upward aim,
With 'jack-in-the-hedge'
To greet my knees;

And shade where stalks proliferate,
Where campions blush,
Where orchids hide:–

Let the airy sounds commingle,
Wind, insect and bird,
And the spent wave:–

Grant me instant oblivion
Of man's drab affairs:–
Channel my thoughts,

Enclose me in far-away mauve,
By the mysteries
Of blurred mountains.

Come, summer, come unsparingly,
Garnish the west wind
With cumulus,

State your terms, flatten the sea, then
Darken it with haze,
Even thunder,

But allow me to remember
Best your finest hour,
Your first deep blue.

Summer Morning
(By the 'Town Farm', Alwington)

Prismatic morning, brilliant, awake
Though not stirring, air soft and reticent,
Sun in orbit, aware of the day's role,
Low-steeped, but generous to trees, grass, shrubs,
Painting life within oak, maple and thorn
As green flushes emerge through their branches,
While dew-soaked verges breathe their shadowed peace
And leaves of nettles and docks irrigate
The high banks where insects patronise the
Cow parsley, majestic, aromatic –
A white cloud against the deepening blue
In this, just the first week of the summer.

Yet only the immediate foreground
Begins to engage in activity
Where, within the 'Town Farm', the milk cooler
Is humming, and queuing cats with arched backs
Guard their cherished positions, facing cows
In long patient rows, three deep, and silent;
But a kenneled dog lets forth angrily,
And birdsong is a mix of syllables
Dominated by insistent pigeons
Who drum out their theme, 'Take *two* cows, Johnny!'
Around swallows content to just hover
Or converse. Aerobatics come later.

Gifts of pure white
(Approaching Alwington parish church)

A bridal gown without need of a showcase,
Summer altarcloth, priestly embroidery,
Fresh compelling incense clinging to the lane,
Dense venerable beard, or cloud inverted
In lattice of white to underpin the sky :–

A jewelled clustering of petals so small
As to be insignificant as flowers,
But, in totality, vast, overwhelming,
Companionable to signposts and farmgates,
Instant hiding place for man or his quarry :–

The uninhibited growth of a few weeks
On seemingly brittle but hardwearing stalks,
Motionless, or impregnated by the wind,
Transforming the green banks – simple cow parsley.

Leisure Pattern
(Westward Ho!)

Sea
Tide out
Ninety-nines
Air-sea rescue
Non-stop cricket score
Dodgems, and all the rest
No fun better than rounders
Non-uniform shapes acquiring
Overall pink, with help from chemist
Moving backpacks on the coastal footpath
Clusters of mermen obsessed by black power
Ocean yachts in advertisements and on tv
But nautical caps may be purchased to boost morale!

Indeterminate!

First you open the windows, then you close them;
You really cannot make up your mind about
Today's weather at all. Now early morning
Was chilly and grey, so that when you went out
You wrapped up well with woollens under your coat,
Only very soon you found yourself too warm,
In fact you were absolutely boiling, thus
You removed the coat – and felt a sudden chill.
And now it is exactly the same indoors:–
Too hot one moment and then freezing the next;
But with the clouds parting and the sight of sun,
Generous sunshine at last, there is no doubt
That the day will be fine and that what you felt,
Or think you felt, earlier was no more than
A hesitant start:– a pity, but this can
Happen sometimes; one must expect ups and downs
Even in full summer, though you forget them,
Smiling, or even singing, with gratitude
As soon as you see the sky turn brilliant.
The only trouble is that, within minutes,
The cloud has returned, the horizon is grey,
It is even raining now, or else thick mist
Is bearing drizzle, but it feels wet enough
As it rushes through the conveniently
Re-opened windows, having soaked the curtains
Already, causing you to resume your rush
Around the house and to offer just the walls,
Or so you hope, the appropriate comment.

Nocturne
(Littleham, Lundy, and Bideford Bay, from Alwington)

Alwington in twilight, its two whitewashed farms
Encase the dim church tower with their icing
In the style of a sepia photograph:–
This is now a hamlet of lemon fingers
Under the parental glare of the full moon,
Spectres that totally refute the lost day's
Pastoral softness with its fusions of green.

Above the church all is spread out prettily
As a map, but easier to recognise,
With hastening night widening the concept,
Not wishing to close it prematurely, and
Permitting clear geographical limits
To be defined by light bulbs:– first, Littleham
Two miles' distant, its hillside surely ablaze
Or else a tilted centenarian's cake:–
Dangerous candles, then, but so compelling
That it would be heartless to puff them all out.

When the vantage point becomes oceanic
The cliffline is visible to Baggy Point,
A greying but still sharp silhouette that clings
To the dim residual light of the sky
While invigorated by the moon-splashed bay,
And eclipsing Saunton's artificial beam
That locates its large rectangular hotel.

Braunton's flares in their wide-set constellation
Play a linking role; some shine out from hollows,
Leaving Georgeham and Heanton, above, as crowns
Perched within their respective bright orange moulds:–

But Lundy's flash, incandescent by design,
And well positioned under a final blush
Of pale rose cast from the sun's dying embers,
Takes on all such challengers, single-handed;
For long hours the island, dimmed by its haze,
Awaited the night to restore what was lost.

Sudden Shower
(Near Beaford Mill)

Poised for attack, an aerial serpent
Who would dispute the sun's authority
To govern, and by clinkering the light
And favouring the arrow's path, not stealth,
To catch his pray, will hiss or even roar
Like the wasting prodigal, or a dam
Without banks, innumerable hoses
Gushing to the sorcerer's command while
Deeply coal-shadowing the valley floor.
But see how every living, bending tree
Defies his artifice, stemming his tide
With the thoroughness of hard metal stakes
Until, worn down, the impact is no more
Than the last puff of a deflated tube;
And the sky, in its sudden cloudless blue,
Admits that natural causes were to blame:—
There was no serpent after all. Then for
The branches, still loaded by the tempest,
Comes their moment to shed all former wrath
And to compete with stone garden fountains,
Brilliant cascades in the Devon sun.

Thunderstorm or Paper Tiger

(A swelling black cloud over Newton Tracey and the memory of a spectre-like tree)

Black your heart, and black your deed, and black the wizened tree you struck–
Terrifying agency, 'Act of God', they say! I doubt it!
But by keeping to terms of meteorology, meaning
That we may profit from our education, the thunderstorm,
Atmospheric instability, heat-triggered or polar,
Is less dreadful when experts reduce it to a diagram!

A Midsummer Day's Dream
(The highways around Barnstaple, and North Devon generally)

Not surprisingly, at the approach to Midsummer Day,
The earlier serenity of the sky was absent
And had been long forgotten. All around me there were folk
Busy dealing with the typical North Devon wetness
Above and on the ground, although had things been otherwise
It would have appeared like an equally typical drought
With all those cheeky little notices – well, are they not? –
Telling us we must 'SAVE WATER NOW!' They were like flowers
That had suddenly sprouted overnight on the verges
Of the highway where advertising is never allowed
Except in dire emergency, which may be often:–

However, enough of that, we have to take the weather
Strictly as it comes and voice our grumbles accordingly,
And because it was ordained that skies should become drear,
And in fact should continue to give a non-stop deluge
Lasting three days and three nights but not forty, thank goodness,
Which is the traditional number needed for a flood,
Just this more modest torrent compared to the Biblical
Catastrophe had everyone screaming with a fervour
That distinguished 1985 from previous years –
No doubt because we had expected Californian
Skies. And now it was agreed that things had gone far enough:–

Except that climatic controls remain tricky affairs
Resistant to desires as well as to tampering,
Making all too pathetic the suffering millions'
Loud protests that they could only stand a certain amount
And that their powers of endurance were far from endless;
Which was the point when I felt bound to ask what were their plans
If the rain did not cease, even then? All contingencies
Should be taken into account, even the likelihood
Of the summer driving folk into their promised frenzy
Once their respective breaking points had been overtaken:–
Just imagine it, the entire population screaming...

And popping off one at a time as ostentatiously
As champagne corks, simply because of inclement weather!
In the meantime the rain continues, and so does my dream
Of what might happen very shortly – any minute now!

After Rain

(South of Torrington, above the Torridge Valley)

Alone, in the quiet places, I begin to live,
Walking the deep lane, with beeches, and the scent of earth,
Where the small echoing sounds are absorbed like a sponge,
And the day's excitable steely showers have gone,
I know not where. Overhead, the cloudy remnants
Are vaguely tinted, and mist creeps over the ground.

Yet my chosen path avoids the saturated air,
There is no flannel to the face, no limpness on clothes;
Rather am I aware of numerous shades of grey
And of the submerging of hedges, gates, fields and trees
That, each for their part, embrace the softness of twilight
And await the anonymity of welcome sleep.

In the distance, where Torrington spreads out like a plain,
The besieging mist, by reinforcing its own strength,
Succeeds in removing most of the frail blurred outlines
Of rows of miniature buildings, and, still extending
Its subtle and ancient blanketwork, converts the town
Into the belly of a large recumbent bear.

Just one unmistakeable landmark resists this change,
The spire of the parish church, an accusingly sharp
Well blackened finger, pointing not so much to the sky
As to a dramatic line of what was once raincloud –
Bulging, still-white cumulus, rising grandly to peaks
That surely await the daring of mountaineers.

For its own part the township now relies totally
On its sodium lights, veritable oranges
Held aloft at differing heights by sticks or wires
That create a paradox:– they are brighter in mist
Than when the night is clear; smudgy, too, the result
Of floating droplets that bend and intensify light.

At first I am mesmerised by this coloured array
And by the dim church with Cromwellian memories,
But as I pick my way homeward, avoiding puddles
That are so obligingly, if feintly, silver-topped,
There are diversions:– static wrought-iron silhouettes
Of cattle against the pale sky, and a rising breeze.

Dream Topping
(From a caravan window at Parklands Farm, Fairy Cross)

It has been a summer of intractable weather, no sun
On demand, yet today the long blades of the field, dew-topped
Before the wind and the sun dry them bear no sign of the past;
Inclemency has been the order of most of their long days,
And while the human spirit marks time, looking for stimulus,
The grass cheerfully gets on with its life and plays its full role.
So why do we shackle oursleves to what are only highlights?
Falling, or rising, the experience can be beautiful:–
All we need is an inner calmness to hear the music that
Will release us from the harmful bondage of always waiting,
Even striving, for wrong things – ephemera, tinted ripples
That we consider essential just because the world does, too.
They may well change surface patterns but not what underlies them,
Yet buying houses, cars, and methods of making life easy,
Wall-to-wall carpets, and kitchens, and attention to food with
The accent on living to eat rather than eating to live
Is becoming an end in itself that insists on the sun
As well, the essential dream topping to sweeten all the gain.
Forget it. Simplicity of life is the sun's best ally;
Everything else is useless vanity, and empties the heart.

Autumn Calm . . .
(Woodtown, and the Yeo Valley)

Days that radiate sun like paintings
In a school exhibition, but mist
Performs its secretive nightly role,
Scooping the valley floor, and rising
As an instant chilly reservoir,
Then vanishing – dissolvable gauze
To allow the light its full passage.

Neighbourly greetings are now proferred
With the calm words of calm inner thought,
And, always, agreeable surprise,
Delight that summer should return while
Suffused with rich Keatsean odours
And a pungency of Brussel sprouts:–
In fact, all scents now cling to the air.

Later the gentility changes
Its form. Blue is replaced by a grey
That is satisfyingly spread out
Across the sky, like a cake filling,
Except that the landscape is now dull,
Shadowless, possibly complacent,
And a dog's bark is two miles' away.

. . . and Gale

(The North Devon coast, the coastal towns, and the hills beyond)

Meteorologists, despite their
Satellites and mathematics should
Look for the swell on the ocean that,
Previously, was beneficent
With its polished metallic surface :–
The first autumn gale has been despatched,
And no seagull will dispute the fact.

The storm may be Icelandic, creeping
Southward, to produce northerly wind,
Or from mid-Atlantic, destined for
The Arctic, either way creating
Foaming hills, ships in acute distress,
And rainfall that pours on to shoppers
Creating havoc with paper bags.

After the continuous passage
Of various technically named
Rainbelts – warm front, cold front, and, later,
The occlusion, which is the merging
Of the two former types into one,
Relief of any kind is welcome,
Even a brighter day with showers.

Yet this may well be one of autumn's
Finest endeavours :– exuberant
Uninhibited blue and sea-green,
Light and shade in a furious chase,
Hills aglow with light and flowing mist,
A vitalising wind in the lungs –
And man shall live for a thousand years!

Who gives thanks for the Harvest?
(Near Hartland Abbey)

The rich man's table never was in doubt:—
His harvest was a matter of degree
In that, should it fail in any year,
He would concern himself most earnestly
With his increasing problems, his estate,
Dowries to be arranged, and whatever
Appeared to be threatened, remembering
His many obligations to the king
In time of war. But now the Crown may rest
While Government insists on what it will:—

Making the formerly rich man humble,
Compelled to share his house with one and all,
While the poor man, knowing he will not starve,
Considers, first, his car and his tv.

Sky Change
(From a cottage window above the Gribble Inn, and outside)

Yes, please do
Be argumentative, insist that one blue sky
Is only marginally different from the next,
Just an expedient gift to enable us
To heave a sigh of relief and part company
With what was half ocean and half blotting paper
Over our head – a precarious existence
Only to be tolerated by those concerned
With bean, pea or cabbage, or whatever it is
That must continue to drink while the sun just waits.

All right, then,
Shall we leave gardening aside for the moment
While we agree that blue skies are manufactured
In the main by high pressure ridges between storms?
Yet I must be free to dwell on their quality,
Their inner essence – mere records are not enough.
How unfortunate you did not witness the dawn
Today, with its brilliant ceremonial
That was too outstanding to be marred by showers.
For once the sky rejoiced in its own perfection

And shone with
A circuitry of cobalt ; it was president
Of the valley mist reinforced by wisps of smoke
But clear of errant haze that saturates the blue
When it bears down with the east wind or the heatwave,
Looking like thundercloud when not concealing it.
On the contrary, this was a day to grow tall,
To feel magnificent, yet the highway ribbon
Still wet from yesterday's rain and whitened by sun,
Like the inn beside it, were nursery products

Made by a
Wizard, because the sky had totally dwarfed them,
Along with the telegraph poles converted to
Metal pins of a black and sinister nature
Beneath receding lines of polished gossamer,
While large-limbed trees withdrew into instant bonzai
Arrangements, and farms were crepe-paper affairs:–
An unfocussed grey sock represented Exmoor,
Though Dartmoor's corrugations well back from the inn
Lay crisp and low at my feet. I could have kicked them!

Frosty Morning
(Woodtown, Fairy Cross)

Branches, twigs, and debris, close
Impacted to the lane by sudden ice,
Mortar of the hard frost, overnight, eventful,
Duly forecast.

Shadow-curve from stockproof fence,
Rime-coated, as the eye investigates,
Probing fields of glitter :— phantom-like, my walk now
Into 'starlight'.

Hedges wearing gaps, with leaves
Obsequious to chill, a galazy
Crinkled into splendour, suffering :— green dampness
Changed by Arctic.

Dimming all that mirage, feint
Opaqueness of the calm, mysterious
Low-born rising finger, viscous tide :— fog driving
Me to firelight.

Concern on Waking
(High on the coastal hinterland)

Successive bursts of raindrops flung
With the insistence of lead shot,
Testing agitated glass panes
To the limits of endurance :–
One can feel tautness of body,
Too, sense imminent disaster,
And then consider how it is
That, if windows endure the blast,
Stoutly built walls cannot escape
Displaying still more sinister
Vibrations? Why be concerned for
A few replaceable items
When total capitulation
Is imminent? Note how the din
Has become a hiss and a roar
From an integration of gusts
That are locked in competition
With each other, not allowing
Perceptible separation
For even the briefest instant.
Later, at the point when prayers
Seem at last to have been answered,
Certainty is not guaranteed :–
One moment the gale is dying
But, the next, it has regained strength ;
And while it all began at night,
Depressing the semi-conscious,
A natural host to terror,
The fantasy is not banished
By the mere rationality
Of waking and seeing the dawn,
For if fear of rampant wind
Is intensified by darkness
And by dream, it is rekindled
By the sight of a cut-throat sky
With its armoury of daggers !

Blizzard!
(Little Torrington, January 17-18, 1985)

Usurper, with wide Cromwellian hat of lead!
Usurper, changed to white, he is the tyrant come
To prove himself with blinding revolution! Yet
He will not tame the green if what will not succumb
Is finer, brighter still when thus it is preserved!

At first the land is difficult to recognise:–
Sufficiency of snow to sugar the cottage,
Gnome-cap the gate, pile pyramid on road, give look
Of far-flung Alpine charm – or else a whitened cage
Endures within the half-stunned focus of both eyes!

Then shall the prisoner discover what he will:–
Exquisite flower of gorse within the weightened hedge –
One daring yellow dart, what impudence! – What next?
My neighbour's wall now biscuit, never grey, and pledge
Of blue through distant mask of haze to saffron hill.

Winter Dawn
(Mainly the River Taw, between Barnstaple and Braunton)

Winter, never termination, but the beginning of opportunity,
And particularly with the calm of a frosty dawn, and the sun fighting
To maintain its command of the sky, achieving most unlikely victory
Against would-be indomitable cloudbanks that appear visibly held back
For a while – one feels sure of it! It is then that the swiftly ebbing channels
Of the Taw and Torridge at low tide, elongated, and contained by mudbanks,
Applaud the gift that is freely offered and, with brilliant alchemy, shed
Eastern fire made more dazzling by the sun's now attenuated angle,
And that can hardly be bettered anywhere on earth – yet unbelievably
Observed when this is the first occasion, and, later, so well remembered as
Outstretched gold wedding rings, superior to metal, and imperishable!

Siberian Link
(North Devon in February 1986)

Not unknown here but no one expected it
To return in any meaningful way, as now,
When, still clutching the old optimistic banner
That proclaims North Devon's Atlantic oasis
Of mildness, barring occasional accidents,
All are suddenly caught off guard and outwitted
By a blast of wind previously known only
To polar bears. My own mind had been centred
On primroses. They would soon be in bloom, I thought,
But, instead, Siberia arrived, either as
Winter's last gamble – or could it be permafrost?

Somehow we must see beyond the present crisis
And the knowledge that one is only half alive,
But thought is arduous and unreliable
Once it has been refrigerated, like the ground
That was once green and the ashen iron-clad sky
Closely linked with the wind, the savage nor'easter,
Militant beyond belief and for that reason,
No doubt, described by Kingsley in his ode as 'brave':–
He had imagination, I will grant him that,
But who in his right mind would welcome a gale
Of ice except in the cause of subterfuge?

The trick must be to tell the hated oppressor
That, on account of his endearing qualities
For which there is always the most widespread demand,
He might care to consider a return visit:–
Perversity would then ensure his refusal.
Another ploy is to humanise the rogue
Element, to place it in line for repentance,
Like the blustering of the west wind followed by
The sun's disarming magic in a poker sky.
There might even be some philosophical use
For the present cold outrage, but it is doubtful.

Remember the Day!

Define the generosity of a day:–
Is it attributable to its sunshine,
To the degree to which temperature soars
Above the normal limit for a season
Noted for black winds and queues at surgeries;
Or to clemency when this is expected
As a matter of course, a virtual right,
Even if it should not be treated as such?
And if it is true that favourable skies
Conspire with expanding earthly bonhomie
To give the human spirit a greater chance
In its struggle to rid itself of despair,
How is it that communities weighed down by
Their daily ephemera, or argument,
Will fail to remember such beneficence
For more than just a minimal length of time?

But the day's rewards defy the popular
Legend that they are directly related
To the benign hand of weather or climate:–
Their occurrence is often unexpected
And may be derived from life's minor events –
Through an open window that generously
Accepts the floral miles of an impish breeze
Or its early morning passage over dew;
Or in stronger wind that is unglamorous
Until it serves to banish mountainous clouds
That for long had cast an unrelieved burden
On the mind. All are generous elements
If they transmute pure chance occurrences
That could so easily have passed unnoticed
Into what may ever after be recalled
And, from each fresh summoning, repeat their joy.

The Constant Vision

They keep us going, do they not? –
Those rosy thoughts of the future
That translate winter dreariness
From a grey illusion of sky
Typical of January
Into the wonder of April's
Freshness and generosity
And the still greater benefits
Through the scented days of May-time
To the hazy calm of the first
Wonderful warmth of the summer:–
Yet when such beneficence fails
Its calendar obligation
There is no need to be fretful,
Merely to project the image
To the day when all will be well.

OTHER POEMS . . .

Where there is Music

Music of the heathered moorland girt by sea,
Music of the sunlit air:– North Devon's fine
Symphonic poem, wind and thunder, ocean's
Rising spray – defiant swell! – yet calm for mind
In sorrow's loss, a Beethoven, for you, for me.

Music of pure birdsong, chaffinch, wren and lark,
Music of the wayout farm – no raucous jolt
To agonise, confront, annoy:– North Devon
Knows her envied role to banish care, support
The blue-backed stars of joy at night, for you, for me.

Music of the breathless common, scented, wild,
Music of the sand and tide, in mirey mouth,
At rosy sundown, or through thicket, urgent
Waterfall, bring images Debussy-wise,
The ever liquid sound of peace, for you, for me.

Music of the manor, music placidly,
Music of the lake and reed; the hollyhock
Of cottage garden, apple, lavender, to
Beech, and larch, and darkened fir where, circling, cry
The buzzard pair in poetry, for you, for me.

Music of the meadow, bee and butterfly,
Music of agility, of swallow's flight
So effortless, so sure, so true – contenders
To the rushing beams of light, and cloud, and shade
That whisper to the heart in dear North Devon land.

Child's first glimpse of the Sea
(Holiday memories of the '30's)

A convincing flash of blue between trees,
Yet held by the eye for only seconds
Before the road teases us with more green
Or descends to jumbled urbanity;
At which point Mother anticipates tea
On arrival – 'hot cups' in her own words –
While no one complains any more of cramp.
Soon the sea is back, bluer than ever,
A curl of smoke on the far horizon
And quickly healing scars from pleasure ships,
Prompting ideas on what we might do:–
There is no shortage of these, for we feel
That time is our servant, like the ocean
Before us and the air that greets our lungs.

Go Calm into the Night

Go calm into the night. Not all can wrestle with miracles,
However great their impact, but they are recognised, and stars
Deserve admiration before mathematics are applied;
And why should heavenly bodies be dulled by discovery? –
For if the moon is intemperate, dusty, and unwatered,
In short, the very essence of grimness, with an unlit sky,
It is different, surely, from ninety-three million miles,
When one may beam back at this wholly obliging extrovert
That will gild vacant air, thoroughly polish all the wet roads,
And entrust dark Delabole slate to join its own shining cause.

Yet, as the moon evades capture in one fascinating glide,
Only the pleated edge of the low cloudbank receives the flare
Of departure, the remainder swells into a black tarnish,
Exemplifying shame in the imperfect workaday world
That now seems less plausible than the Milky Way, or Venus,
Or the well-worn Plough with one fading but still tenacious lamp –
A mischievous fellow judging him by his knowing twinkle
Among brighter stalwarts who are totally predictable:–
Observe the sky how you will, hear the cries of far galaxies
That, for all their junketing, are bound, too, by limitations.

There is just one thing. If heat and light obey given command,
Be that from God, or the laws of energy, or both at once,
Are there no accidents above, events defying control,
Spelling the finite chink in envisaged perpetuity –
The nebular equivalent to the human buffer-car
Existence that must be played out by knocking into others?
At least, the very distance of all possible worlds in space
Is great enough to be unchallenging to those who love night
For its peace, grieving not for the day until it reclaims us
And we resume our role. Will there be miracles here as well?

Call o' the North

Devon, you called
I came to you,
Weary of town,
Dreaming one thing,
Finding it true.

Ocean so wide,
Windy and blue,
Foaming in white,
Cry o' the gull,
North in my sight.

Rugged your shore,
Hearty your men,
Millers in vale,
Sailors at sea –
God protect them.

Take winter's pow'r,
Thunder on rock,
Yours is the law,
Wayward your blast,
North to the core.

Breeze by the hill,
All season through,
Bright after storm,
Calm after gale,
No one like you.

Mist below cloud,
Valleys in sun,
Wonderful land,
Tang o' the air,
North took my hand.

Rivers below
Forests of pine,
Widely you flow,
Secrets you hold,
Yet you are mine.

Moorland of mauve,
Banks of deep green,
Golden the beach,
Silver the rain,
Sweetly, you teach.

Cows giving milk,
Busy the farm,
Dairy for cream,
Swarming of bees,
Owls in the barn.

North is so free,
North made the call,
North land alone,
Restless old North,
North to the bone.

With acknowledgment to John Galsworthy's
'Devon for Me'